THE POETRY OF GERMANIUM

The Poetry of Germanium

Walter the Educator™

SKB

Silent King Books a WhichHead imprint

Copyright © 2023 by Walter the Educator™

All rights reserved. No part of this book may be reproduced in any manner whatsoever without written permission except in the case of brief quotations embodied in critical articles and reviews.

First Printing, 2023

Disclaimer
This book is a literary work; poems are not about specific persons, locations, situations, and/or circumstances unless mentioned in a historical context. This book is for entertainment and informational purposes only. The author and publisher offer this information without warranties expressed or implied. No matter the grounds, neither the author nor the publisher will be accountable for any losses, injuries, or other damages caused by the reader's use of this book. The use of this book acknowledges an understanding and acceptance of this disclaimer.

"Earning a degree in chemistry changed my life!"
— Walter the Educator

dedicated to all the chemistry lovers, like myself, across the world

CONTENTS

Dedication v

Why I Created This Book? 1

One - Symbol Of Innovation 2

Two - The Messenger 4

Three - A Gem 6

Four - Silent Hero 8

Five - Germanium's Might 10

Six - Stellar Element 12

Seven - Oh Germanium 14

Eight - Day By Day 16

Nine - Stands As A Catalyst 18

Ten - Power Flows 20

Eleven - Progress And Shine 22

Twelve - Element Rare 24

Thirteen - Brilliance Shines	. . . .	26
Fourteen - Symbol Of Ingenuity	. . . .	28
Fifteen - Hidden Secrets	. . . .	30
Sixteen - Breaking Boundaries	. . . .	32
Seventeen - Perfect Rhyme	. . . .	34
Eighteen - Hour After Golden Hour	. . . . .	36
Nineteen - The Conductor	. . . .	38
Twenty - Thought And Sound	. . . .	40
Twenty-One - Connects The Dots	. . . . . .	42
Twenty-Two - Silent Partner	. . . .	44
Twenty-Three - Transistors To Diodes	. . .	46
Twenty-Four - Unlock The Mysteries	. . . .	48
Twenty-Five - Gatekeeper Of Progress	. . .	50
Twenty-Six - Wondrous Element	. . . .	52
Twenty-Seven - Forever Draped	. . . .	54
Twenty-Eight - Element Profound	. . . . . .	56
Twenty-Nine - Galaxies Afar	. . . .	58
Thirty - Ever-bright	. . . .	60
Thirty-One - Nothing To Hide	. . . .	62
Thirty-Two - Brighter Day	. . . .	64

Thirty-Three - Dreams Shall Stay 66

Thirty-Four - Germanium's Essence 68

Thirty-Five - Vastness Of Space 70

Thirty-Six - Celebrated Eternally 72

About The Author 74

WHY I CREATED THIS BOOK?

Creating a poetry book about the chemical element Germanium was an intriguing and unique endeavor. Germanium, with its atomic number 32, has several interesting properties that can inspire poetic themes and metaphors. Its association with semiconductors, its role in the development of technology, and its connection to the periodic table can provide a rich backdrop for exploring various themes such as transformation, resilience, and the intersection of science and art. By combining scientific knowledge with the creative power of poetry, this book can offer a fresh and captivating perspective on the world of chemistry and inspire readers to see beauty in unexpected places.

ONE

SYMBOL OF INNOVATION

In the realm of elements, Germanium shines,
A crystal so rare, its beauty defines.
A metalloid it is, with atomic weight,
Symbol Ge, it holds a mysterious fate.

In the depths of Earth, where minerals lie,
Germanium forms, unseen by the eye.
A treasure concealed, its secrets untold,
A story of wonder, waiting to unfold.

With properties unique, it captures the light,
Conducting electricity, oh what a sight!
An element of promise, in the world of tech,
Germanium's prowess, we cannot neglect.

From fiber optics to semiconductors,
Germanium's role, a conductor of whispers.

It fuels our devices, with power and speed,
A silent hero, fulfilling our every need.

But Germanium's reach extends far and wide,
In health and science, it takes us for a ride.
Used in infrared technology, it sees beyond,
Revealing secrets, the unseen, it's bond.

Oh Germanium, element of intrigue,
Your presence in nature, a true masterpiece.
A testament to the wonders of creation,
Germanium, a symbol of innovation.

TWO

THE MESSENGER

In the realm of elements, a hidden gem shines bright,
Germanium, a conductor of electrons, a wondrous sight.
With atomic number thirty-two, it holds its ground,
A symbol of innovation, in science it is found.

In the depths of the periodic table, it lies,
A silent force, with secrets it implies.
Its crystalline structure, a marvel to behold,
A testament to nature's mysteries untold.

Germanium, oh marvel, with properties sublime,
A bridge between metals and non-metals, so fine.
Through fiber optics, it carries light's embrace,
Connecting the world with its ethereal grace.

In the realm of semiconductors, it finds its home,
Unlocking the power of technology, a stepping stone.

With its ability to conduct, yet resist with might,
Germanium paves the way, shining so bright.

    Infrared technology, it embraces with zest,
Seeing beyond the visible, the unseen, the best.
A window to a world, where vision extends,
Germanium, the messenger, that technology sends.

    So let us celebrate this element divine,
A hidden gem, a treasure of science, so fine.
Germanium, oh wonder, with endless possibility,
In the realm of elements, you shine with superiority.

THREE

A GEM

In the realm of science, a gem so rare,
Germanium shines with brilliance and flair.
A conductor of electrons, it plays its part,
Guiding innovation, igniting the spark.

With atomic number thirty-two it stands,
In the periodic table, among noble bands.
Its crystal structure, a marvel to behold,
A lustrous gem, with stories yet untold.

Germanium, the semiconductor's delight,
In technology's realm, it takes flight.
From transistors to fiber optics, it weaves,
A tapestry of progress with every reprieve.

Infrared detectors, oh how they rely,
On Germanium's prowess, they amplify.

With thermal imaging, it reveals the unseen,
Unveiling secrets, in the dark, serene.

Its thermal conductivity, a treasure to find,
Dissipating heat, with a purposeful mind.
A boon to electronics, it never fails,
Keeping devices cool, as innovation prevails.

So let us celebrate this element rare,
Germanium, a gem beyond compare.
For in its atomic structure, we find,
The power to shape the future, unconfined.

FOUR

SILENT HERO

In the depths of the periodic table,
A gem unseen, its story untold,
Germanium, a metalloid so stable,
A treasure to behold.

A shimmering crystal, it takes its form,
With lustrous hues of gray,
A conductor, a semi, it can transform,
In technological play.

In the realm of transistors and wires,
Germanium finds its place,
Its conductivity never tires,
Advancing with steady pace.

Within the realm of infrared,
Germanium holds its sway,

As photons dance, its path is led,
In the darkness, it holds the ray.

With its semiconducting skill,
It paves the way for innovation,
From smartphones to solar panels, it fills,
The world with its transformation.

Oh Germanium, the silent hero,
Dissipating heat with grace,
In circuits, its power does grow,
As technology finds its space.

A marvel of the periodic table,
Germanium, embraced by our hands,
In each device, it is stable,
A symbol of progress that expands.

FIVE

GERMANIUM'S MIGHT

In the realm where science meets art,
Lies a gem that plays its part.
Germanium, a metalloid so rare,
With properties beyond compare.

Amidst the periodic table it resides,
A conductor, a semiconductor it hides.
With four valence electrons in its core,
It weaves a tale of innovation galore.

Germanium, oh versatile element,
In the world of technology, you're heaven-sent.
Your crystal structure, oh so neat,
Forms a lattice, a symphony complete.

As a thermal conductor, you excel,
In electronics, your secrets you tell.

From transistors to fiber optics, you shine,
Connecting the world, one line at a time.

Germanium, a trailblazer in its own right,
A catalyst for progress, shining so bright.
From computers to solar cells, you're the key,
Empowering the future, for all to see.

So let us raise a toast to Germanium's might,
A vital ingredient in our digital flight.
In the realm where science meets art,
Germanium, you'll forever hold a special part.

SIX

STELLAR ELEMENT

In the realm of the periodic table,
Where elements entwine and enable,
Lies a gem in the atomic sea,
A treasure known as Germanium, you see.
    With a shimmering hue, it stands,
A conductor of nature's finest strands,
Its valiant electrons dance and sway,
In the world of technology, they hold sway.
    Germanium, oh noble element,
A symbol of advancement, a testament,
To the wonders of science and innovation,
Fueling progress, a catalyst without hesitation.
    In the depths of transistors, it dwells,
A vital component, where magic compels,

In circuits and chips, its presence is found,
Enabling communication without a sound.

    From computers to smartphones, it plays a part,
Connecting the world, with a beating heart,
A bridge between minds, a global thread,
In the digital age, it's widely spread.

    Germanium, a conductor of light,
In fiber optics, it shines bright,
Guiding information at lightning speed,
Through oceans and skies, it takes the lead.

    So let us praise this stellar element,
For its contributions, so truly evident,
In technology's realm, it reigns supreme,
Germanium, the catalyst of our wildest dreams.

SEVEN

OH GERMANIUM

In the realm of technology, a gem shines bright,
A lustrous element, casting its radiant light.
Germanium, the marvel, with secrets untold,
Unveiling wonders, as its story unfolds.

Through fiber optics, it weaves a web so fine,
Transmitting data, across the cosmic line.
A conductor of dreams, it beckons us all,
To a world of connectivity, where boundaries fall.

In semiconductors, it finds its true place,
Guiding the flow, with precision and grace.
An electron dance, orchestrated so well,
Germanium's magic, in each transistor dwell.

Infrared technology, it brings warmth to the night,
Revealing the unseen, with its thermal sight.

A guardian of secrets, it pierces the dark,
Germanium's vision, leaves an indelible mark.

Oh Germanium, element of might,
Your essence exudes, a luminous light.
A silent hero, in the world of tech,
Advancing progress, with every single speck.

So let us celebrate, this element grand,
For its contributions, to the digital land.
Germanium, the catalyst, that propels us ahead,
Innovation and wonder, on its path we tread.

EIGHT

DAY BY DAY

In the realm of technology, Germanium shines,
A precious element, a gem so fine.
Its thermal conductivity, a gift to behold,
Connecting the world through fibers of gold.

Through the web of communication, it weaves,
A symphony of data, the world receives.
With speed and precision, it carries our voice,
Germanium, the conductor of our choice.

Infrared technology, it harnesses with might,
Revealing the unseen, in the darkest of night.
Through its lenses, the hidden world is revealed,
A glimpse into mysteries, forever concealed.

But Germanium's contributions don't end there,
In electronics, its presence is rare.

A semiconducting marvel, it paves the way,
For progress and innovation, day by day.
   In transistors, it dances, with grace and ease,
Connecting the world, as we please.
Through wires and signals, it breathes life anew,
Germanium, the catalyst, that makes dreams come true.
   Oh, Germanium, you are a force to admire,
With your magic and might, you never tire.
In the realm of technology, you stand tall,
Advancing progress, inspiring us all.

NINE

STANDS AS A CATALYST

In the realm of technology, Germanium stands tall,
A shining element, a marvel to us all.
With thermal conductivity, it dissipates the heat,
Advancing progress, innovation at its feet.

From transistors to diodes, it plays a vital role,
In circuits and devices, where it takes control.
Its semiconducting abilities, a gift to behold,
In the world of electronics, it's worth more than gold.

Germanium, oh Germanium, with fibers it does dance,
Guiding information, with elegance and chance.
In the realm of fiber optics, it leads the way,
Whispering secrets, transmitting light's array.

Infrared technology, where the unseen is revealed,
Germanium unlocks the hidden, forever unconcealed.

Peering through the darkness, it shows us the way,
Revealing the mysteries, where shadows often stray.
And in the realm of progress, where innovation thrives,
Germanium stands as a catalyst, the element that drives.
Electronics owe their existence, to its noble form,
A symbol of ingenuity, through the calm and storm.
Germanium, oh Germanium, a marvel to behold,
In the world of elements, your worth is untold.
A pioneer of technology, a beacon of light,
Germanium, oh Germanium, forever shining bright.

TEN

POWER FLOWS

In the realm of technology, it finds its place,
A lustrous element, with a conductive embrace.
Germanium, the connector of our world,
A symbol of innovation, beautifully unfurled.

In the transistor's realm, it plays its part,
A tiny hero, igniting progress like a spark.
With its semiconducting prowess, it shines,
Unleashing a revolution, with boundaries left behind.

From infrared technology, its power flows,
Revealing the unseen, with every radiant glow.
Through the depths of darkness, it guides our way,
A beacon of light, leading us astray.

Oh, Germanium, a catalyst for change,
A symbol of ingenuity, so wondrously strange.

From its humble origins, to its present might,
It fuels our advancements, with endless delight.
    So let us celebrate this element rare,
With gratitude and awe, let's all share.
For Germanium, in all its splendid grace,
Connects our world, and lights up the space.

ELEVEN

PROGRESS AND SHINE

In realms of silicon, where circuits reside,
A wondrous element emerges, side by side.
Germanium, with its crystalline grace,
An enigma of progress, a marvel to embrace.

In technology's realm, it finds its place,
Connecting the world with electronic grace.
A semiconductor, it paves the way,
Unleashing innovations day by day.

A catalyst for change, it sparks the fire,
Igniting progress, taking us higher.
From transistors to fiber-optic strands,
Germanium's touch, a magician's hands.

Its valence electrons dance in delight,
Navigating pathways, shining so bright.

A conductor in disguise, it defies,
Guiding the flow, where progress lies.
   Germanium, the unseen revealer,
Infrared's ally, a truth concealer.
Unveiling the secrets, the hidden truth,
Through its lens, unveiling youth.
   In the world of elements, it stands unique,
A symbol of ingenuity, a treasure to seek.
Germanium, oh element divine,
Your presence, a catalyst for progress and shine.

TWELVE

ELEMENT RARE

In the heart of the periodic table's realm,
Lies a gem, a treasure, Germanium.
A metalloid with powers untold,
Unveiling secrets, mysteries unfold.
  In the realm of technology it resides,
Where progress and innovation coincide.
Semiconducting marvel, it takes the lead,
Bringing forth the future with remarkable speed.
  With silicon companions, it forms a bond,
Creating circuits, connections beyond.
Transistors and diodes, the building blocks,
Powering the world with electric shocks.
  A catalyst for progress, it takes its place,
Igniting the flames of the human race.

Infrared wavelengths dance upon its face,
Revealing the unseen, with elegance and grace.

From fiber optics to solar cells,
Germanium's presence, a tale it tells.
It lights up the night, it powers the day,
A guiding force, showing us the way.

Oh Germanium, element divine,
Your contributions forever shine.
In the realms of science and technology,
You are the spark, the source of creativity.

So let us celebrate this element rare,
For it is Germanium that we must share.
A symbol of progress, a symbol of might,
Germanium, the beacon of light.

THIRTEEN

BRILLIANCE SHINES

In the realm of technology's embrace,
A gem of the elements, with grace,
Germanium, the unseen's revealer,
A catalyst for change, an innovator.

Beneath the surface, it quietly gleams,
Unearthing secrets, fulfilling dreams,
Its atomic structure, a marvel to behold,
A story of progress, yet untold.

Infrared waves, they dance and sway,
Germanium's touch, they obey,
Guiding us through the darkness of night,
Revealing hidden wonders, pure delight.

A conductor of electrons, it plays its part,
A bridge between worlds, a work of art,
Transistors and diodes, its humble abode,
Creating pathways, where ideas explode.

From fiber optics to solar cells,
Germanium's touch, in each tale it tells,
A pioneer of progress, it leads the way,
A beacon of hope, in every display.

Oh, Germanium, your brilliance shines,
Through circuits and screens, your legacy climbs,
Forever ingrained in the fabric of time,
A symbol of innovation, sublime.

So let us celebrate, this element divine,
Germanium, the unseen's design,
For in its embrace, we find our stride,
Guided by its light, we reach new heights.

FOURTEEN

SYMBOL OF INGENUITY

In the realm of progress, a catalyst so grand,
Germanium, the element, takes a stand.
A conductor of electrons, a guide to the light,
It sparks innovation, with its power so bright.

In the world of technology, it finds its place,
Germanium's presence, a celestial grace.
From transistors to diodes, it paves the way,
Unleashing the future, every single day.

With silicon and germanium, a bond so strong,
They form the backbone, where dreams belong.
In circuits and chips, they dance and they play,
Creating a symphony, in a digital array.

Germanium, the unseen, it reveals the way,
Guiding us through darkness, night and day.

Infrared eyes, it lends us its sight,
Unveiling the secrets, hidden from light.

Advancing progress, inspiring us all,
Germanium's essence, we proudly extol.
Through fiber optics it travels, so swift,
Connecting the world, with a magical lift.

Solar cells harness its energy divine,
Harbingers of change, in a world so fine.
Germanium, the conductor, the catalyst true,
Igniting the spark, in everything we do.

So let us celebrate, this element rare,
A symbol of ingenuity, beyond compare.
Germanium, the beacon, shining so bright,
Guiding our journey, towards a future so right.

FIFTEEN

HIDDEN SECRETS

In the realm of elements, Germanium stands,
A catalyst for progress, in inventive hands.
With atomic number thirty-two, it resides,
A symbol of innovation, where knowledge resides.

Conductor of electrons, it bridges the divide,
Between worlds of science, where dreams coincide.
Infrared technology, its domain so vast,
Unveiling hidden secrets, from the past.

Fiber optics it embraces, with grace and flair,
Transmitting data swiftly, through strands so rare.
Solar cells it energizes, with radiant light,
Harnessing the power of the sun, shining so bright.

Germanium, oh brilliance, guide us through the dark,
Illuminate the mysteries, where shadows embark.
A symbol of creativity, in its purest form,

Inspiring minds to flourish, and boundaries to transform.

With every discovery, a new chapter unfolds,
Germanium's brilliance, forever enrolls.
A beacon in the night, it leads us towards,
A future filled with wonder, where progress soars.

So let us celebrate, this element divine,
Germanium, the catalyst, that forever will shine.
Innovation and brilliance, its legacy will be,
A symbol of human potential, for all to see.

SIXTEEN

BREAKING BOUNDARIES

Germanium, the catalyst of change,
A humble element, yet so profound.
A conductor of both heat and sound,
Its impact on progress, nothing short of range.

Innovation, its power to evoke,
A bridge between worlds, a connector supreme.
Reveals the unseen, like a lucid dream,
Its ability to guide us through darkness, it's no joke.

Technology, its realm of expertise,
A vital component in the modern age.
A conductor, it paves the way to new stages,
Its significance, a key ingredient to progress with ease.

Germanium, the beacon of hope,
A catalyst for solar energy, so bright.

Inspiring creativity, breaking boundaries with might,
A reminder to never stop, to never elope.

In this world of constant change and flux,
Germanium, the element that connects us.

SEVENTEEN

PERFECT RHYME

In the realm of progress, a catalyst we find,
A gleaming element with a brilliant mind.
Germanium, the conductor of dreams,
Unveils the unseen, guiding us through streams.

    A crystal lattice, elegant and strong,
A beacon of knowledge, where wonders belong.
Its atomic structure, a symphony of grace,
Unleashing potential, in every place.

    In the realm of technology, it takes its stand,
A cornerstone of innovation, hand in hand.
From transistors to optical fibers so fine,
Germanium's essence, a tapestry divine.

    It beckons us forward, to the future unknown,
With its power to transform, to build and to own.

In the depths of darkness, it dances with light,
A guiding force, igniting our sight.
    Germanium, the guardian of progress and more,
Unleashing ideas, like never before.
Through its conductivity, electrons dance,
Unlocking the secrets of chance.
    So let us celebrate this element grand,
With its power to shape, to understand.
Germanium, the catalyst of our time,
A symphony of atoms, in perfect rhyme.

EIGHTEEN

HOUR AFTER GOLDEN HOUR

In the realm of innovation, a conductor of might,
Germanium, the element, shining ever bright.
A catalyst for progress, a marvel to behold,
Unleashing boundless wonders, its story yet untold.

With its crystalline lattice, its structure so refined,
Germanium paves the way, leading humankind.
In the realm of technology, it takes center stage,
Empowering our devices, igniting the digital age.

From transistors to semiconductors, it weaves its magic spell,
Guiding electrons with precision, where data stories dwell.
In the heart of every circuit, it pulses, it thrives,
An elixir of connectivity, sparking dreams alive.

Germanium, oh Germanium, your brilliance knows

no bounds,
You amplify our voices, in whispers and in sounds.
A conductor of harmony, in the symphony of light,
Revealing hidden wonders, painting colors in the night.

Oh, Germanium, you are the solar ray, the source of boundless power,
Harnessing the sun's energy, hour after golden hour.
You pave the way for progress, in a world that seeks to change,
A beacon of innovation, in a world that's rearranged.

So, let us celebrate Germanium, its wonders, and its might,
A catalyst for progress, shining ever bright.
In labs and in our hearts, its legacy will thrive,
Guiding us towards a future where possibilities will thrive.

NINETEEN

THE CONDUCTOR

In the realm of atoms, a gem does reside,
A conductor of electrons, Germanium's pride.
With a crystal lattice, it shines so bright,
Guiding us through darkness, like a beacon of light.

In fiber optics, it weaves a tale,
Transmitting data, without ever to fail.
Through thin strands of glass, it carries the load,
Connecting the world, with a digital ode.

Solar cells harness its power so pure,
Capturing sunlight, a gift to endure.
Transforming photons into energy sublime,
Germanium's brilliance, a limitless climb.

In labs and in circuits, it plays its part,
Amplifying signals, with an electric art.

A catalyst for progress, it sparks innovation,
Unlocking secrets, with each new revelation.

Oh, Germanium, element of wonder,
You inspire us all, like a rolling thunder.
With your atomic number of thirty-two,
You pave the way for a future anew.

So let us embrace your metallic grace,
And explore the depths of your electron chase.
For in your essence, we find endless possibilities,
Germanium, the conductor of our technological realities.

TWENTY

THOUGHT AND SOUND

In the realm where atoms dance and collide,
There lies a gem, a secret to confide.
Germanium, a metalloid of grace,
A conductor of light, a celestial embrace.

In solar cells, it finds its noble role,
Harnessing the sun, a power to extol.
Through crystal lattice, photons intertwine,
Generating energy, a gift divine.

Oh, Germanium, conductor of progress and light,
Catalyst of innovation, shining so bright.
In circuits and chips, you pave the way,
A symphony of electrons, in harmony they sway.

Through darkness, you guide us with steadfast might,
Illuminating paths, casting away the night.

In fibers and lenses, your brilliance we find,
Revealing hidden truths, expanding our mind.

In the world of technology, you reign supreme,
Enabling communication, like a lucid dream.
Transistors and diodes, with your guiding hand,
Unleashing the potential of a digital band.

Oh, Germanium, amplifier of thought and sound,
Capturing voices, spreading them around.
In cameras and microphones, your artistry unfolds,
Preserving memories, like stories yet untold.

Versatile and precious, you hold the key,
Unlocking mysteries, inspiring creativity.
Germanium, symbol of human progress and might,
With you, we soar, reaching new heights.

TWENTY-ONE

CONNECTS THE DOTS

In the realm of solar might, a gem does shine so bright,
Germanium, the element, harnessing the sun's pure light.
A semiconductor, with powers yet untold,
It transforms photons into energy, a tale to behold.

Bathed in the sun's golden blaze, Germanium awakens,
Absorbing and converting, as its purpose beckons.
In solar cells, it plays a crucial role,
Capturing the sun's essence, igniting our souls.

With every ray that touches its surface, a dance begins,
Electrons flow, creating a current that never dims.
From dawn till dusk, it tirelessly works,

Unleashing the sun's potential, breaking free from the murk.

But Germanium's reach extends beyond solar's domain,
In the realm of technology, it sets the world aflame.
A conductor of innovation, it connects the dots,
Building bridges between minds, where creativity trots.

In transistors and diodes, it paves the way,
Enabling progress, day by day.
With its unique properties, it lights the path,
Empowering the future with an innovative wrath.

Oh, Germanium, conductor of dreams,
In your atomic structure, progress gleams.
From solar cells to technology's might,
You amplify our hopes, shining ever so bright.

TWENTY-TWO

SILENT PARTNER

In the realm of science, where secrets lie,
There shines a metal, with grace, it's called Germanium.
A conductor of dreams, a catalyst of change,
It weaves through the cosmos, in a celestial range.

With atomic number thirty-two, it claims its place,
An element so rare, yet it leaves no trace.
Its crystalline structure, a sight to behold,
Whispering secrets, as stories unfold.

Germanium, the amplifier of light,
In solar cells, it shines ever so bright.
Harnessing energy from the sun's gentle rays,
Powering the world, igniting new ways.

In the world of technology, it finds its might,
A bridge between circuits, a beacon of light.

Connecting the dots, unveiling the unknown,
Unleashing innovation, where it's been sown.

Germanium, oh conductor of dreams,
Revealing hidden truths, or so it seems.
Through its metallic grace, it paves the way,
For progress and discovery, day by day.

So let us embrace this shimmering element,
With potential untold, and dreams we invent.
For Germanium's brilliance, it knows no bounds,
A silent partner, in scientific grounds.

TWENTY-THREE

TRANSISTORS TO DIODES

In the realm of circuits, where energy flows,
A metalloid emerges, its power it bestows.
Germanium, the element, with atomic might,
Harnessing the sun's rays, it shines so bright.

In solar cells, it captures the light's embrace,
Transforming photons with elegant grace.
A conductor of dreams, it connects the way,
Unveiling new horizons, where progress may lay.

Oh, Germanium, you unlock hidden truths,
Revealing pathways to innovation's roots.
In the realm of technology, you take your stand,
Empowering devices with a gentle hand.

From transistors to diodes, you pave the way,
Enabling communication, day after day.

In circuits, you dance with precision and grace,
A vital component in this digital space.
    Oh, Germanium, you are more than a mere element,
A catalyst for progress, a profound testament.
In labs and workshops, your secrets unfold,
Inspiring minds, as stories of wonder are told.
    So, let us celebrate this versatile gem,
Germanium, the element, a true diadem.
For in your presence, we find endless worth,
A symphony of atoms, a masterpiece of Earth.

TWENTY-FOUR

UNLOCK THE MYSTERIES

In the realm of atoms, a treasure I find,
A shimmering gem of a different kind.
Germanium, they call you, a metalloid rare,
With secrets and wonders beyond compare.

In solar cells, your brilliance shines bright,
Harnessing energy from the sun's pure light.
A conductor of power, you pave the way,
For technology's progress, day after day.

You bridge the gap, between circuits and more,
A conductor of innovation, at the core.
With your presence, the signals flow,
Unlocking the potential, that lies below.

Germanium, you hold the key,
To reveal hidden truths, for all to see.

In semiconductors, you play your part,
Advancements in science, you help impart.

From transistors to fiber optics so grand,
You enhance communication across the land.
In our modern world, you leave your mark,
A catalyst for progress, a shining spark.

Oh, Germanium, element of might,
Your versatility brings us delight.
From technology's realms to scientific domains,
You shape our world, where progress reigns.

So, let us celebrate, this element divine,
Germanium, a treasure we'll forever enshrine.
For in your essence, we see the power,
To unlock the mysteries, hour after hour.

TWENTY-FIVE

GATEKEEPER OF PROGRESS

In the realm of sun-kissed dreams, behold Germanium's gleam,
A jewel of science, a catalyst unseen.
Within the realm of atoms, it finds its sacred place,
A conduit of power, a beacon of grace.

In solar cells it dances, harnessing the sun's embrace,
Transforming light to energy, with elegance and grace.
It weaves a tapestry of hope, where innovation thrives,
Empowering a world where every dream survives.

Germanium, oh Germanium, a guardian of the new,
A conductor of knowledge, connecting me and you.
Through fibers thin as whispers, it carries our desires,
Uniting hearts and minds, igniting passion's fires.

In transistors it resides, a gatekeeper of progress,
Innovations bloom, as its secrets we possess.
From computers to smartphones, it shapes our digital age,
A symphony of circuits, where possibilities engage.

Germanium, oh Germanium, a chameleon of might,
Adapting to the needs, in every field, shining bright.
Infrared spectrometers, it reveals hidden truths,
Unraveling the mysteries, exposing nature's cues.

So let us raise our voices, in praise of Germanium's might,
A humble element, yet a beacon of light.
It binds us all together, in a world forever changed,
Germanium, oh Germanium, forever we are engaged.

TWENTY-SIX

WONDROUS ELEMENT

In the realm of science and innovation,
Lies a metalloid of great fascination,
Germanium, a gem with a shimmering sheen,
Unveiling secrets, a technological dream.

Within the heart of the digital age,
Germanium takes center stage,
A semiconductor, it sparks the fire,
Igniting progress with its endless desire.

From transistors to fiber-optic strands,
Germanium's versatility expands,
In communication's vast domain,
It weaves a web of signals, a vibrant chain.

Through its lens, mysteries are unlocked,
In laboratories, minds are truly shocked,
Its properties, a scientist's delight,
Germanium's magic shines so bright.

It binds electrons, forming pathways clear,
Carrying information, far and near,
In the realm of quantum mechanics,
Germanium dances, defying limits.

Oh, Germanium, you shape our world,
With you, innovation is unfurled,
From computers to smartphones in our hands,
You're the key to progress, the digital band.

So, let us celebrate this wondrous element,
Its impact on technology, truly eminent,
Germanium, a symbol of human might,
Guiding us towards a future so bright.

TWENTY-SEVEN

FOREVER DRAPED

In the realm of technology's might,
There lies a gem, a hidden light.
Germanium, the element of power,
Embodied in every passing hour.

Within the circuits, it finds its place,
A conductor of electronic grace.
Transistors dance with perfect precision,
Unlocking realms of communication.

From computers to phones, it weaves its spell,
Connecting souls from afar to dwell.
Through optic fibers, it carries the words,
Unveiling secrets, like fluttering birds.

Oh, Germanium, your versatility shines,
A beacon of progress, a symbol divine.

Infrared detectors, you do create,
Revealing truths in a world so great.
   In nature's lab, you were formed,
A testament to progress, ever transformed.
With each innovation, you pave the way,
For a future where nothing will ever sway.
   Germanium, you carry the code,
Unleashing the power to shape the abode.
Through your properties, the world is shaped,
A legacy of knowledge, forever draped.

TWENTY-EIGHT

ELEMENT PROFOUND

In the realm of technology, a marvel yet unseen,
Lies Germanium, a gem of silicon's kin.
With valence electrons, its lattice takes form,
A conductor, a semiconductor, through which signals transform.

In the heart of the transistor, it finds its domain,
A switch and amplifier, a power to sustain.
From integrated circuits to microchips, it thrives,
Bridging the gap between digital lives.

Resistant to heat, a thermal conductor indeed,
Germanium paves the way for progress to heed.
Infrared detectors and fiber-optic strands,
Its light-sensitive nature, our world expands.

In the depths of the cosmos, it seeks to explore,
X-ray detectors, spectrometers, unlocking the core.

From distant galaxies to atomic design,
Germanium's insight, a treasure we find.
   Boundless in its potential, this element profound,
Carrying the promise of knowledge unbound.
Technology's ally, a bridge we traverse,
Germanium's symphony, our world diverse.

TWENTY-NINE

GALAXIES AFAR

In the realm of technology's embrace,
There lies a gem that's full of grace.
Germanium, the element of might,
Unveiling secrets, shining bright.

With crystal lattice, pure and strong,
It sings the song of the digital throng.
Transistors whisper, circuits hum,
Germanium's magic, a symphony strum.

From telegraph wires to fiber-optic dreams,
Germanium's light, a messenger it seems.
Through signals it carries, information flows,
Connecting the world, wherever it goes.

In quantum realms, it takes its flight,
Germanium's versatility, a wondrous sight.

Adaptable and potent, it dances with grace,
Unraveling mysteries, in quantum's embrace.
 As a conductor of heat, it stands with pride,
Defying the flames, it won't subside.
In furnaces it thrives, under pressure it shines,
Germanium's resilience, a testament to find.
 Across the cosmos, it boldly ventures,
Exploring the universe, as it enters.
From distant planets to galaxies afar,
Germanium's spirit, a guiding star.
 So let us honor this element divine,
For Germanium's legacy will forever shine.
In technology's tapestry, it weaves its thread,
Bringing progress and wonder, where it's led.

THIRTY

EVER-BRIGHT

In the realm of elements, let us praise Germanium,
A versatile gem, shining with brilliance,
Amidst the periodic table's vast dominion,
It stands, a symbol of scientific resilience.

With atomic number thirty-two, it resides,
A bridge between metals and nonmetals,
Its properties, a marvel, none can deride,
As Germanium's story gracefully unfolds.

In technology's realm, it finds its stage,
A semiconductor, with electrons dancing,
A key ingredient, on the modern page,
Transforming our world, advancing.

Through heat and pressure, it stands unyielding,
A testament to its strength, unwavering,
From transistors to fiber optics, revealing,
The secrets of progress, forever unswerving.

In quantum mechanics, it takes the lead,
Unraveling mysteries, beyond our sight,
Unlocking the door, where wonders breed,
Guiding us towards the future, shining bright.

A thermal conductor, brave and true,
In flames and pressure, it never bends,
Exploring the cosmos, as if on cue,
Revealing the universe, where it transcends.

Oh, Germanium, we honor your legacy,
A catalyst for technology's grand parade,
From science to progress, you've set us free,
In your presence, we find solace, unswayed.

So let us celebrate this element divine,
Germanium, a beacon of knowledge's might,
In every field, its wonders intertwine,
Guiding us towards a future, ever-bright.

THIRTY-ONE

NOTHING TO HIDE

In the realm of elements, a wonder we find,
Germanium, a gem, both rare and refined.
A versatile soul, with secrets untold,
Unfolding mysteries, in its atoms, unfold.

In technology's realm, it takes center stage,
With circuits and chips, it writes a new page.
A conductor of power, it paves the way,
For innovation and progress, day by day.

Oh, Germanium! Your resilience shines bright,
Exploring the cosmos, reaching for the light.
Through telescopes vast, you gaze at the stars,
Unlocking the universe, removing its bars.

You guide us towards a future, bold and bright,
With your presence, we'll conquer any plight.

As a thermal conductor, you keep us in balance,
In the world of heat, you offer your guidance.

Unraveling mysteries, hidden from our sight,
Germanium, you lead us with your cosmic light.
From the depths of the Earth, to the expanse of space,
Your legacy, eternal, no one can erase.

In our devices, you'll forever reside,
A symbol of progress, with nothing to hide.
Germanium, we salute your noble quest,
With gratitude and awe, we hold you to our chest.

THIRTY-TWO

BRIGHTER DAY

In the realm of elements, a gem so rare,
Germanium, a wonder beyond compare.
With versatility it reigns supreme,
A symbol of resilience, a dreamer's theme.
 From Earth's crust, it emerges bright,
A catalyst for progress, a source of light.
In technology's embrace, it finds its place,
Enabling innovation, at a rapid pace.
 A conductor of electrons, it does excel,
Through heat and power, its story to tell.
In the cosmos, it ventures wide and far,
Guiding us towards a future, like a guiding star.
 From telescopes that pierce the night,
To satellites in orbit, a celestial sight.

Germanium's presence, a legacy profound,
Unlocking mysteries, where knowledge is found.
   With each discovery, a step towards the unknown,
A journey of wonder, a seedling sown.
Germanium, the pathfinder of our kind,
Leading us towards a future, bold and bright.
   Embrace its power, its atomic might,
Let it ignite your curiosity, day and night.
For in Germanium's realm, we find our way,
To a world of progress and a brighter day.
   So let us honor this element so grand,
With gratitude and awe, let us take a stand.
For Germanium's legacy shall forever endure,
A beacon of inspiration, forever pure.

THIRTY-THREE

DREAMS SHALL STAY

In the realm of elements, Germanium stands tall,
A versatile gem, that surpasses them all.
Resilient and strong, it weathers the test,
With properties unique, and a brilliance expressed.
    In technology's grasp, Germanium thrives,
A conductor of progress, where innovation thrives.
In transistors and diodes, its power unfolds,
As it paves the way for a future untold.
    In exploration's quest, Germanium guides,
Unraveling mysteries, where the universe hides.
In telescopes and sensors, it takes the lead,
Revealing secrets, the cosmos decreed.
    From fiber optics to solar cells, it imparts,
A warmth in its heart, that connects us in parts.

With thermal conductivity, it dissipates heat,
Unlocking new frontiers, where knowledge shall meet.

In the fields of medicine, Germanium imparts,
Its healing touch, mending broken hearts.
With its presence profound, it aids in the fight,
Against ailments and diseases, bringing hope and light.

Germanium, oh Germanium, a beacon so bright,
Leading us forward, with its radiant light.
In satellites soaring, it guides our way,
Towards a brighter future, where dreams shall stay.

So let us celebrate, this element divine,
Germanium's glory, forever shall shine.
A testament to human ingenuity and grace,
Germanium, the jewel, in the elements' embrace.

THIRTY-FOUR

GERMANIUM'S ESSENCE

In the realm of technology, bright and bold,
There lies a metal that's worth its weight in gold.
Germanium, the conductor of dreams,
Ignites innovation, like sun's fiery beams.

Through circuits and wires, it weaves its spell,
Guiding electrons with a magical swell.
Resilient and steadfast, it stands the test,
A foundation of progress, the very best.

In the depths of the cosmos, it searches afar,
Exploring the universe, a celestial star.
Germanium telescopes, reaching for the sky,
Unveiling mysteries, as time passes by.

But not only in space does its power reside,
In medicine's realm, it offers a guide.

A healer, a helper, a beacon of light,
Germanium fights ailments, with all its might.
    Conducting heat, it spreads warmth and grace,
Embracing the world in its warm embrace.
A symbol of hope, inspiration, and more,
Germanium's essence, forever we adore.
    So let us celebrate this element divine,
For its versatility, its brilliance does shine.
Germanium, the catalyst of our time,
We salute you, for you are truly sublime.

THIRTY-FIVE

VASTNESS OF SPACE

In the realm of atoms and cosmic dust,
There lies a gem, rare and robust.
Germanium, oh element of might,
Unveiling the cosmos, day and night.

Through the galaxies it gracefully soars,
Unraveling mysteries, unlocking doors.
Its electrons dance in a symphony,
Guiding us on an extraordinary journey.

From Earth's crust to the stars above,
Germanium shines with cosmic love.
Its crystalline lattice, pure and strong,
A symphony of atoms, a celestial song.

Thermal conductivity, its wondrous art,
Leading us toward a brighter start.
Harnessing heat, dissipating flame,
Germanium, a beacon of innovation's flame.

In medicine's embrace, it brings us healing,
A balm for ailments, a gentle feeling.
Healing wounds, soothing the soul,
Germanium's magic, making us whole.

And in the lenses of telescopic sight,
Germanium reveals secrets, day and night.
Peering into the vastness of space,
Uncovering wonders, with elegance and grace.

Germanium, a symbol of progress and might,
A catalyst for dreams, burning bright.
In science, medicine, and the cosmos vast,
Forever inspiring, as time flies fast.

THIRTY-SIX

CELEBRATED, ETERNALLY

Germanium, oh element divine,
With mysteries hidden, yet to shine.
In the depths of science, you reside,
Guiding us towards a brighter stride.

A semiconductor, you stand tall,
A catalyst for progress, above all.
Your atomic structure, so precise,
Unlocks the secrets, a priceless device.

Through lenses of Germanium, we see,
The wonders of the universe, set free.
You amplify signals, with utmost care,
A conduit of knowledge, beyond compare.

In circuits and transistors, you excel,
A conductor of energy, we can tell.

Thermal conductivity, your gift,
Dissipating heat, a seamless lift.

Germanium, you heal and restore,
Infrared rays, a remedy to explore.
With wavelengths soothing, you mend,
A symbol of healing, till the end.

In space exploration, you take flight,
Radiation detectors, shining bright.
Unraveling the cosmos, far and wide,
Germanium, our guide, with cosmic pride.

Versatile in nature, you adapt,
In fiber optics, you're firmly strapped.
From medicine to technology's boast,
Germanium, you inspire, utmost.

Oh Germanium, element of might,
A beacon of progress, shining bright.
Innovation's ally, you will be,
Forever celebrated, eternally.

ABOUT THE AUTHOR

Walter the Educator is one of the pseudonyms for Walter Anderson. Formally educated in Chemistry, Business, and Education, he is an educator, an author, a diverse entrepreneur, and he is the son of a disabled war veteran. "Walter the Educator" shares his time between educating and creating. He holds interests and owns several creative projects that entertain, enlighten, enhance, and educate, hoping to inspire and motivate you.

Follow, find new works, and stay up to date
with Walter the Educator™
at WaltertheEducator.com

www.ingramcontent.com/pod-product-compliance
Lightning Source LLC
LaVergne TN
LVHW051958060526
838201LV00059B/3714